Hugo Mauricio Jiménez M.
Silvia Rosy Gómez D.

Manual of Laboratory Practices in Biotechnology

Hugo Mauricio Jiménez M.
Silvia Rosy Gómez D.

Manual of Laboratory Practices in Biotechnology

Industrial Biotechnology, Environmental Biotechnology and Molecular Biology Laboratory Guides

ScienciaScripts

Imprint

Any brand names and product names mentioned in this book are subject to trademark, brand or patent protection and are trademarks or registered trademarks of their respective holders. The use of brand names, product names, common names, trade names, product descriptions etc. even without a particular marking in this work is in no way to be construed to mean that such names may be regarded as unrestricted in respect of trademark and brand protection legislation and could thus be used by anyone.

Cover image: www.ingimage.com

This book is a translation from the original published under ISBN 978-620-2-81400-3.

Publisher:
Sciencia Scripts
is a trademark of
International Book Market Service Ltd., member of OmniScriptum Publishing Group
17 Meldrum Street, Beau Bassin 71504, Mauritius
Printed at: see last page
ISBN: 978-620-3-16285-1

Biotechnology Laboratory Practice Manual.

By:

Prof. Hugo Mauricio Jimenez M.
Microbiologist, M. Se.

Prof. Silvia Rosy Gómez D.
Bacteriologist, M. Se.

November 2020.

Content

Presentation

Microbiology as a recent multidisciplinary science that integrates biotechnological advances in areas of study such as medicine, environment and industry, has allowed the development of various microbiological products of biotechnological interest of application such as biological control, bioremediation and large-scale production of antibiotics, vaccines, anticancer drugs, enzymes, bioinsecticides and biofertilizers, among others.

It is important to understand biotechnology from the biological, environmental and industrial points of view, which would contribute to the development of a scientific capacity in students of Biological Sciences.

Likewise, on our planet Earth there is a great biodiversity; an estimated 15 million species have been taxonomically classified into the main groups; viruses, archaeobacteria, bacteria, fungi, protozoa, algae, plants, nematodes, mollusks, crustaceans, insects, arachnids, birds, amphibians, reptiles, fish and mammals.

In Colombia a mega-diverse country due to its variability of ecosystems, climate, soil and having the Atlantic and Pacific oceans on its coasts, has a high diversity of birds, butterflies, orchids, plants, reptiles, amphibians and primates, and about the microbial diversity is very little known, for this reason it is important to conduct studies that lead to the knowledge of this diversity as it would be the application of microorganisms such as bacteria and microfungi for biotechnological purposes.

Bioprospecting is defined as the search for organisms such as archaeobacteria, bacteria, fungi, algae, plants and animals of benefit to mankind from a Medical, Environmental or Agronomic point of view.

With the development of these Practical Biotechnology Laboratory Guides presented in this Manual, students will be able to integrate other areas of knowledge such as industrial and environmental microbiology, microbial morphology and physiology, molecular biology, biochemistry, biophysics, among others, and will also acquire skills and abilities in the handling of laboratory instruments and equipment and manipulation of microorganisms such as bacteria and microfungi, and also develop cognitive skills through the methodology learning - doing, and generate teaching strategies for the study of the physiology and biology of bacteria and microfungi, which encourages attitudinal and scientific learning directed towards research in Biotechnology, Industrial Microbiology, Environmental Microbiology, and Molecular Biology.
This type of cognitive and attitudinal learning generates in the students research skills and a positive attitude, which allows an intellectual, scientific and ethical development; this type of teaching in integral formative research in the students will be of great contribution in their future professional life.

This Biotechnology Laboratory Practice Manual contains 10 Practical Guides designed from the professional experiences of the authors and the application of bacteria and microfungi and are easy to carry out with the reagents and equipment of a basic biology laboratory.

These 10 Practical Laboratory Guides serve as scientific training for students and professionals in the Biological Sciences, and also as motivation for the design, development and implementation of research projects in Biotechnology.

We hope that this Manual of Biotechnology Laboratory Practices is to your liking and very useful in your research work.

Hugo Mauricio Jimenez M.

Introduction

There are many definitions of Biotechnology and one of them is that of the Convention on Biological Diversity (United Nations, 1992) which indicates that it is *"any technological application that uses* biological *systems and living organisms or their derivatives to create or modify products or processes for specific uses,* which is called "Modern Biotechnology" but those that have been performed for thousands of years as the processes of fermentation by microorganisms for the production of foods such as beer, wine, bread, cheese, and yogurt, among others, which are grouped in the "Traditional Biotechnology" (Occelli, 2013).

As can be seen, biotechnology has been present throughout the history of humanity with a social, environmental and economic impact since its inception, and it is still a constituent element of the culture of the 21st century. In Colombia, the National Biotechnology Program contributes to the *"increase of development, welfare and economic competitiveness from the knowledge, protection and use of the* (Minciencias, 2020).

Therefore, every organism has a biotechnological potential with influence in the agricultural, environmental, medical and industrial fields, among others. In addition, biotechnology as a multidisciplinary science contributes to the apprehension of updated disciplinary contents related to biodiversity, as well as to the strengthening and development of scientific, attitudinal, procedural, evaluative and metacognitive competences.

The objective of the "Manual of Laboratory Practices in Biotechnology" is to create a research culture in Biotechnology by showing easy methodologies through the use of bacteria and microfungi. It contains 10 Practical Laboratory Guides that are made up of: introduction, objectives, materials and reagents, methodology, questionnaire and bibliography.

We believe that this material will not only contribute to the scientific training of students and professionals but also to the motivation for the design, development and implementation of research projects in various areas of biotechnology.

In the area of Industrial Biotechnology, the Practical Laboratory Guides: *Acetic* Acid Production by *Acetobacter* sp and *Gluconobacter* sp bacteria, Acid-Lactic Fermentation: Yogurt Production and Alcoholic Fermentation: Wine Production, *In Vitro* Screening of Cellulases from *Penicillium aurantiogriseum, In Vitro* Amylase Production Tests with *Aspergillus fumigatus,* will allow the reader to know, learn and apply microorganisms of interest in fermentation and enzyme production.
In Environmental Biotechnology the Practical Laboratory Guides: Crude Oil Biodegradation Bioassays with *Pseudomonas fluorescens,* Antagonistic Potential of *Trichoderma harzianum,* Biofertilizers for the Improvement of Soy Growth and Development, will stimulate the reader to know and learn about topics related to Bioremediation, Biological Control and Biofertilizers.
In Molecular Biology the Practical Laboratory Guides: Extraction of nuclear DNA from *Saccharomyces cerevisiae* and Extraction of plasmid DNA from *Pseudomonas fluorescens* will allow the reader to learn about techniques for extracting DNA from

microorganisms.

Therefore, our main objective of this Manual of Laboratory Practices in Biotechnology is to create a research culture in Biotechnology showing some methodologies through the application of bacteria and microfungi, and that are easy to perform and can be taken to various areas of research.

Bibliography

Minciences. National Program of Biotechnology of Colciencias. 2020 http://www.colciencias.gov.co/node/1133

United Nations. 1992. *Convention on Biological Diversity.* Rio de Janeiro-Brazil: United Nations. Available at: http://www.cbd.int/convention/articles/?a=cbd-02

Occelli, M. 2013. Teaching biotechnology in school: contributions and didactic reflections. Biological Bulletin Magazine n° 27 - year 7 P. 9-13.

Production of *Acetic* Acid by the bacteria *Acetobacter* sp and *Gluconobacter* sp.

By: Hugo Mauricio Jimenez M.

Introduction

Acetic acid, also called ethanoic acid or methylenecarboxylic acid, is an organic acid with two carbon atoms, and can be found in the form of acetate ion. Its formula is CH3-COOH (C2H4O2), being the carboxyl group the one that gives the acid properties to the molecule. This is an acid found in vinegar, which produces its sour taste and smell (Speight, James G. 2002).

Acetic acid is produced by the fermentation of various substrates, such as starch solution, sugar solutions, or alcoholic food products such as wine or cider, by means of acetic acid bacteria. This acetic acid is obtained by synthesis and by bacterial fermentation, which contributes 10% of the world's production. The 75% obtained in the chemical industry is prepared by methanol carbonation (Speight, James G. 2002).

Acetic acid bacteria are naturally present in the grapes, they can withstand the conditions of vinification and reach the maturation of the wine with an active metabolism, these bacteria are great producers of acetic acid and acetaldehyde, raising the volatile acidity of the wine and need oxygen to grow; the processes that oxygenate the wine favor the growth and metabolism of these bacteria (Hernández, I & f. Barbero 2008).

Acetic acid bacteria carry out the acetification of fermented products due to their ability to oxidize alcohol to acetic acid. These bacteria are important in the wine industry because they can alter the organoleptic characteristics and quality of wine by oxidizing ethyl alcohol to acetic acid.

Acetic acid bacteria (BAA) belong to the family *Acetobacteriaceae;* they are included in the group of a-Proteobacteria. They are Gram-negative microorganisms, ellipsoidal or cylindrical in shape that can be found isolated, in pairs or forming chains. They are mobile by polar or peric flagellation. They present positive catalase activity, negative oxidase and do not form endospores. They use oxygen as the final acceptor of electrons, so they have a strict aerobic metabolism, with oxygen as the final acceptor of electrons (Gerard, L. 2015).
The oxidation of ethanol to acetic acid is the best known characteristic of acetic acid bacteria, this biochemical process consists of two stages: in the first the ethanol is transformed into acetaldehyde by the enzyme alcohol dehydrogenase (ADH) and then, the acetaldehyde is transformed into acetic acid by the enzyme acetaldehyde dehydrogenase (ALDH) (Gerard, L. 2015).

Some bacteria can produce high concentrations of acetic acid, up to 50g/L (Gerard, L. 2015), this characteristic is very important for the vinegar industry.

Acetic acid bacteria genera such as *Acetobacter, Gluconobacter and Gluconacetobacter* can be used for the development of a culture for the oxidation of alcoholic musts obtained from fruits, in order to obtain acetic acid (Gerard, L. 2015). At an industrial level, acetic

"

acid bacteria have an industrial importance in the production of vinegar (acetic acid).

The transformation of pineapple juice into acetic acid is a biotechnological process which is produced by two successive fermentations: alcoholic and acetic.

In alcoholic fermentation, Saccharomyces *cerevisiae* yeast transforms sucrose into ethanol, an anaerobic process which is more efficient due to the biochemical characteristics of the pineapple. Then the *Acetobacter* sp and *Gluconobacter* sp bacteria oxidize the ethanol and produce acetic acid, through an aerobic process.

According to General Microbiology, 2008-2009, cited by Bernal, C & L. Cortes, 2010) other methods of acetic acid production are:

Orleans Method: it is carried out by filling the fourth part of a wooden barrel used for the maturation of the wine with fresh vinegar obtained by fermentation of *Acetobacter* sp and *Gluconobacter* sp which provides a fresh inoculum, then the fermented alcoholic drink is added, the container is left open so that an exchange of oxygen takes place, the process lasts several weeks and the efficiency depends on the availability of oxygen.

Bubbling Method: This is a submerged fermentation process in which oxygen is supplied through an air bubbling process. The speed of ethanol addition is regulated to allow efficient conversion to vinegar, reaching a production of 98%.

Some applications of acetic acid according to www.ecured.cu/Acido_acetico are

- Used as a condiment
- It is used in the manufacture of esters or essences.
- Color fixer
- Solvent
- Raw material in obtaining acetone, acetates, aspirin and other derivatives
- In beekeeping it is used to control the larvae and eggs of wax moths.
- Production of sodium acetate and as an antibiotic extraction agent.
- As a bactericide.
- Neutralizer and in dyeing processes in the textile and leather industry.
- As an acidifying agent and for the preparation of fruit esters in the food industry.
- Insecticide ingredient.

Objectives

- Perform a Fermentation for ethanol production from fermented pineapple juice using *Saccharomyces cerevisiae* yeast.

- Observe the ethanol production by *Saccharomyces cerevisiae.*

- To carry out a Fermentation for acetic acid production from fermented pineapple juice using *Acetobacter* sp and *Gluconobacter* sp bacteria.

- Observe the production of acetic acid by *Acetobacter* sp and *Gluconobacter* sp.

- Develop skills in the handling of laboratory instruments

Materials, reagents and equipment.

- Active yeast of yeast: *Saccharomyces cerevisiae.*

- Pure cultures of *Acetobacter* sp and *Gluconobacter* sp.

- Water distills sterile.
- Asa
- Dark Flasks.
- Sucrose.
- pH tape.
- Pineapple juice.
- Scale.
- Breathalyzer.
- cork or gauze and cotton stopper

Ethanol Production

Methodology:

1. In dark bottles add a liter of pineapple juice, then add 5 g of active yeast of levapan and 10 g of sucrose, shake well, cover with cork (or gauze and cotton stopper) and leave at an average temperature of 20 °C for 7 days.

2. After 7 days, place the 1000 mL of the culture in a 1000 mL test tube and with an alcoholometer measure the percentage of ethanol. If the ethanol production is low, leave another 7 days and measure again the ethanol percentage.

3. Each time you perform step 2, make the respective tasting, bouquet and texture of the liquor.

Acetic Acid Production:

1. In the dark bottles of one liter where the alcoholic fermentation was carried out (ethanol production), add 10 mL of inoculum* of the bacteria *Acetobacter* sp and *Gluconobacter* sp, shake well, cover with cork (or gauze and cotton stopper) and leave at an average temperature of 20 °C for 7 days.

*This inoculum is obtained by adding 5 mL of sterile distilled water in fresh cultures respectively of *Acetobacter* sp and *Gluconobacter* sp in inclined test tubes with Nutritional Agar detaching with round handle.

2. After the 7 days put in the pH measurement with a pH tape, and report the result.

3. Carry out the respective tasting, if it presents vinegar flavor, the result is positive.

Questionnaire

1. Explain the metabolic pathway of ethanol production through glycolysis.

2. Explain the metabolic pathway of ethanol oxidation for acetic acid production.
3. Is this bioprocess for acetic acid production on a large or small scale? Explain why.
4. How would you do a Bioprocess scaling for acetic acid production?

5. See the scientific name of 10 bacteria related to acetic acid production.

Bibliography

- Cindy Bernal, Lina Cortes, 2010.
Isolation, Characterization and Conservation of Acidic - Acetic Bacteria from traditional fermented products. Grade work. Department of Biology - UPN. Colombia.

- Cindy Bernal, Lina Cortes, Hugo Mauricio Jimenez M. 2011.
Isolation, Characterization and Conservation of Acidic - Acetic Bacteria from traditional fermented products as a pedagogical tool.
BIO - GRAPHICS. Vol 4, No. 7-2011., pp. 108-111. ISSN 2027 - 1034.

- Gerard, L. 2015. Characterization of acetic acid bacteria for the production of fruit vinegars. Doctoral Thesis National University of Entre Rios and Polytechnic University of Valencia. Available at: https://riunet.upv.es/bitstream/handle/10251/59401/GERARD Review Date: 06/03/2019.

- Hernández, I & f. Barbero, 2008. Acetic acid bacteria: detection and elimination techniques.
Available at:
http://www.guserbiot.com/pdf/Guserbiot_Viticultura_Bacterias_Aceticas.pdf
Revision Date: 07/03/2019.

- General Microbiology. 2008 - 2009. Genetics and Microbiology Research Group: http://www.unavarra.es/genmic/hall/docencia.htm, cited by Cindy Bernal, Lina Cortes, 2010.

- Speight, James G. 2002. Chemical Process and Design Handbook. McGraw-Hill Cited in: Acetic Acid: Available at: www.ecured.cu/Acido_acetico. Revision date: 05/03/2019.

Acid-lactic fermentation: Yogurt production

By: Hugo Mauricio Jimenez M.

Introduction

Yogurt is a very old food. The first traces of its existence date back to between 10,000 and 5,000 BC, in the Neolithic period.

The origin of yogurt is located in Turkey although there are also those who locate it in the Balkan Peninsula, Bulgaria or Central Asia. Its name comes from a Bulgarian term, iaurt. It is believed that its consumption predates the beginning of agriculture.

The nomadic peoples transported the fresh milk they obtained from the animals in sacks generally made of goatskin. The heat and the contact of the milk with the skin of goat propitiated the multiplication of the acid bacteria that fermented the milk. The milk became a semi-solid and coagulated mass. Once the lactic ferment contained in those bags was consumed, they were filled again with fresh milk that was transformed again into fermented milk by the residues that remained.

Yogurt became the basic food of nomadic peoples because of its ease of transport and conservation. Its healthy virtues were already known in antiquity.

Yogurt is a form of modified sour milk. For its elaboration, it can be made not only from cow's milk but also from goat's and sheep's milk, whole, partial or totally skimmed, previously boiled or pasteurized.

The type of milk used for its elaboration depends on the place where it is elaborated and consumed. In Central, North and South America, as well as in Western Europe the preference and production is based on cow's milk; in Turkey and Eastern Europe on goat's milk and in Egypt and India on buffalo's milk.

Today, yogurt is widely recognized as a healthy food. Manufacturers have responded to the growth in yogurt consumption by introducing many different types of yogurt, including low-fat and 0%, creamy, liquid to drink, organic, baby, with fruit and ice cream. The basic ingredients and their manufacture are practically similar:

- First, the raw milk is transported from the farm to the factory, where it will be processed.
- When the milk arrives at the plant, its composition is modified before being used to make yogurt. The milk is then standardized by its dry extract, pasteurized (at 80°C) and homogenized.
- Once the pasteurization and homogenization processes are completed, the milk has to be cooled to 43-46 °C and the fermentation culture added at a concentration of about 2 %. The cultures are composed of two lactic acid bacteria: *Streptococcus thermophilus and Lactobacillus delbrueckii subsp. bulgaricus,* and *Lactococcus lactis.* These bacteria ferment their consistency, taste, aroma and health benefits, as well as facilitating digestion.
- After cooling, fruit, sugar and other ingredients can be added to obtain a wide variety of products, and then the yogurt is packaged.

- Finally, the product is cooled and stored at refrigeration temperatures (4 °C) for preservation.

Types of yogurt:

- In the beaten yogurt, the milk is fermented in a fermentation tank with a coating. After fermentation, the contents are mixed, and fruits and aromas are added. It is then left to cool and the products are packaged and stored at refrigeration temperatures.
- In the firm yogurt, also known as French-style, the milk is inoculated with ferments and other ingredients (fruit preparation, sugar, aromas) are added before packaging. The fermentation process takes place in the containers during the incubation period, once the product has been cooled and stored at refrigeration temperatures.
- Liquid yogurt is a beaten yogurt that has a low total solids content and is subjected to a homogenization process to reduce its viscosity. Subsequently, sweetening, aromatic or coloring ingredients can be added and finally the product is packaged in bottles.

During the lactic fermentation (yogurt production), the milk sugar (lactose) is first transformed into simple sugars, specifically glucose and galactose, and then converted into lactic acid, this provides an acidity (pH 4.5), which precipitates the proteins (caseins) and concentrates the milk which gives the yogurt its special texture, thus creating the specific texture of yogurt, in addition, the lactic fermentation produces peptides, amino acids that give the yogurt its characteristic flavor.

The nutrient composition of yogurt is based on the nutrient composition of the milk from which it is derived. The final composition is determined by the source and type of milk solids added prior to fermentation, the lactic fermentation and bacteria strains used in the fermentation, the temperature, the duration of the fermentation process, the storage time, and the ingredients (such as fruit) that may be added to the yogurt.

13

Benefits of Yogurt: It is a source of protein and milk fat, due to the bacterial fermentation process, lactose is also fermented, which means that people intolerant to lactose can consume yogurt. In addition, it is rich in calcium and some B vitamins. It is also beneficial to the immune system, because it helps fight infections and reduces the negative effects of antibiotics. In addition, it stabilizes the intestinal flora and the set of microorganisms of the digestive system. It favors the absorption of fats, so it is an ally against overweight, is good for the skin and fights diarrhea and constipation, as well as facilitates the assimilation of nutrients and reduces cholesterol.

Objectives

- Performing an acid-lactic fermentation from milk using a culture of
Lactobacillus bulgaricus, Lactococcus lactis, and *Streptococcus thermophilus*

- Obtain the ability to analyze a Lactic Fermentation.

- Develop skills in the handling of laboratory instruments

- Potentiate cognitive skills.

Materials, reagents and equipment.

- 1 Litre of milk.
- Thermometer.
- Powdered milk.
- A large spoon.
- Yogurt without candy (cultivation).
- Fruit or jam.
- Plastic cups.
- Refrigerator.

General Yogurt Production Process:

1. Selection of the milk: the milk must have a high protein content.

2. Pasteurization: the purpose is to concentrate the serum proteins, the ideal temperature is 88 °C to 95 °C for 5 to 10 minutes.
3. Concentration: powdered milk is added, this is dissolved in warm milk and then yogurt is added.

4. Culture: composed of acidic-lactic bacteria: *Lactobacillus bulgaricus* and *Lactococcus lactis* The culture is incubated for 2 hours at 44 °C.

5. Sowing: after pasteurization the milk is cooled and the crop is sown in the proportion of 3% by shaking well.

6. Packaging: it is sown in plastic cups and covered with vinipel.

7. Incubation: it is incubated at a temperature of 42 °C.

8. Yogurt preparation: fruits or jams are added after incubation.

9. Conservation: left in the fridge at 4 °C.

Methodology

1. Boil 1 liter of milk and cool to 60°C.

2. Add 3 tablespoons of powdered milk, mix and cool to 42 °C.

3. Add 3 tablespoons of yogurt without candy (*Lactobacillus bulgaricus* and *Lactococcus lactis* culture) and mix.

4. Add fruit or jam.

5. Serve in plastic bottles.

6. Incubate at 44 °C for 4 hours.

7. Refrigerate at 4 °C.

8. Taste, 12 hours after refrigeration

Questionnaire

1. Explain the metabolic pathway of lactic acid production through glycolysis.

2. See the scientific name of 3 other bacteria used in the production of yogurt.

3. How would you do a Bioprocess scaling for industrial yogurt production?
4. Consult companies dedicated to the production of yogurt in Colombia.

Bibliography

- Lozada, K. 2000. Industrial Microbiology Laboratory Guides. University of the Andes. Science Faculty. Department of Biological Sciences. Colombia.

Cybergraphy

- What is Yogurt, available at: https://ww-w.yogurtinnutrition.com/es/que-es-el-yogur-FAQs/ Revision Date: 12/03/2019

- The yogurt, available at: https://www.zonadiet.com/bebidas/yogurt.htm Review Date: 12/03/2019

- Yogurt, available at https://es.wikipedia.Org/wiki/Yogur#cite_note-monografia-7. Revision Date: 12/03/2019.

- Natural Yogurt, available at https://biotrendies.com/lacteos/yogur-natural. Revision: 12/03/2019.

Alcoholic Fermentation: Wine Production.

By: Hugo Mauricio Jimenez M.

Introduction

History

The cultivation of the vine (*Vitis vinifera sylvestris*) and the production of drinks from grapes (in the form of juices with added sugar) were already being carried out around 6,000 and 5,000 BC, but it is not until the Bronze Age (3,000 BC) that the real birth of wine is estimated to have taken place. Archaeologists have found evidence that establishes the origin of the first wine harvest in Sumer, in the fertile lands of the Tigris and Euphrates in ancient Mesopotamia.

From Sumer it arrived in Egypt, where it would rival the beer that was brewed in Ancient Egypt (3,000 B.C.). The banks of the Nile were lands of cultivation of the vine and in volume to these plants, a whole labor and industrial activity was developed. The Egyptians fermented the must in large earthenware vessels and produced red wine. The wine became a symbol of social status and was used in religious rites and pagan festivities. The pharaohs were buried with earthenware vessels containing wine and engravings have been found on the pyramids symbolizing the cultivation of the vine, the harvesting, elaboration and enjoyment of wine at festivals and religious events.

The adaptability of the vine (*Vitis vinifera)* favoured its expansion throughout Western Europe through trade routes, reaching as far as China. It is believed that the vine reached the Iberian Peninsula before the Phoenicians around 3,000 BC.

In 700 B.C., the wine arrives in its expansion process to the classic Greece. The Greeks drank wine in religious rites, funerals and popular festivals. They also assigned a deity to wine: Dyonysos, who is always represented with a glass in his hand. The Greeks created containers of different sizes for the storage of wine: large *amphorae,* which were sealed with pine resin; medium sized *craters;* and small *aoinojé* and *rhytons.*

Wine production:

The type of wine is mainly determined by the variety of grape, its secondary metabolites are the source of the aroma, color and flavor of the wine. Its concentration depends on the variety, the microclimate and the grape plant. The vine requires low nitrogenous nutrition. An excess of nitrogenous fertilizers can be negative for the taste components of the wine (Wyss, G & Bon van Elzakker, 2005).

According to the International Wine Campus, the parts of the wine production process are

-Harvest: is the collection of the grape, for example in Spain is done between the months of September and October. In addition, when the grapes are collected they must show a suitable state of ripeness in order to extract the best quality from them.

- **Destemming:** in this process the grapes are separated from the rest of the bunch, the objective being to separate the grapes from the branches and/or leaves because they provide flavors and aromas that are bitter for the production of wine.

- **Crushing:** the grapes are passed through a treading machine to get the grape skin, called *skin, to be* broken. This way the juice is extracted to make the next step easier, but it should not be crushed too much to avoid breaking the seeds of the grapes, which would bring bitterness to the wine.

- **Maceration** and **fermentation:** The juice that is extracted will be kept at a controlled temperature for a few days, thus allowing the fermentation and thus acquiring the required color. In these tanks and through its own yeast, the alcoholic fermentation process begins as the sugar in the grapes ends up being transformed into ethyl alcohol. This process lasts, depending on the type of wine and must take place at temperatures not exceeding 29°C.

- **Pressing:** as the solid product of fermentation still contains large quantities of wine after *devatting* (action that consists of separating the wine from the solid parts of the grape), it is subjected to a pressing to extract the liquid.

- **Malolactic fermentation:** the wine obtained during the previous steps is put through a new fermentation process. Through this process the acid character of the wine is reduced and it becomes much more pleasant to drink.

- **Ageing: the** process of maturing, ageing or aging is one of the most important points for the elaboration of a wine. In this process, the wine is introduced in barrels so that it acquires aromatic characteristics that can be distinguished during tasting. In the barrels, the wine evolves and develops. While the wine matures in the barrels, two additional tasks are carried out to eliminate impurities and sediments such as *racking and* clarification.

- **Bottling:** the wine evolves and assimilates the oxygen that is introduced into the bottle. **Alcoholic fermentation** is the process by which a yeast, in this case *Saccharomyces cerevisiae,* produces ethanol from a carbon source, with glucose, fructose and sucrose being the most widely used, under conditions of low oxygenation, average temperature of 20 °C and no light (Jimenez, H. 2013).

Alcoholic beverages can be produced from different substrates such as the juices of different fruits enriched with sucrose through a small-scale fermentation process. During the process, time must be standardized, since over long periods ethanol converts to acetic acid producing acidity and damaging the texture and flavor of the liquor (Jimenez, H. 2013).

Saccharomyces cerevisiae yeast is generally osmophilic, i.e. it resists high concentrations of sugar and tolerates high concentrations of ethanol, about 20%.

Precisely, some limitations of the Bioprocess is the high concentration of ethanol, today by means of the improvement of strains are obtaining yeasts that resist concentrations of ethanol higher than 20%. Another limitation is the pH, values lower than 3.5 decrease the production of ethanol, for this reason it is important not to use acid fruits in these processes. The high concentration of sugars decreases the efficiency of fermentation, aeration increases yeast cell respiration and diverts the ethanol production process to non-fermentative vegetative growth.

Liquid biofuels such as bioethanol, which is a main derivative of sugarcane fermentation, are generally considered a sustainable solution to energy and environmental problems (Jie et al, 2012).

In Bioprocesses for the production of bioethanol, the scale-up begins with fermentations by lot (Batch) and then fermentation by fed lot (Fed Batch), to then take it to Pilot Plant and once standardized take it to commercial fermentations (500,000 liters).

Objectives

- Perform an alcoholic Fermentation from different substrates (fruit juices) using *Saccharomyces cerevisiae* yeast.

- Obtain the ability to analyze an Alcoholic Fermentation

- Develop skills in the handling of laboratory instruments
- Potentiate cognitive and scientific skills.

Materials, reagents and equipment

- Active yeast of yeast *(Saccharomyces cerevisicie)*.
- Dark Flasks.
- Sucrose.
- pH tape.
- Pineapple juice.
- Scale.
- Breathalyzer.
- cork or gauze and cotton stopper

Methodology

1. In dark bottles add a liter of fruit juice of pifia or red grape, then add 5 g of active yeast of levapan and two spoonfuls of sugar (sucrose), shake well, cover with cork (or gauze and cotton stopper) and leave at an average temperature of 20 °C for 7 days.

2. After 7 days, place the 1000 mL of the culture in a 1000 mL test tube and with an alcoholometer measure the percentage of ethanol. If the ethanol production is low, leave another 7 days and measure again the ethanol percentage.

3. Each time you perform step 2, make the respective tasting, bouquet and texture of the liquor.

Questionnaire

1. Explain the metabolic pathway of ethanol production through glycolysis.

2. Explain the importance of ethanol.

3. How would you do a bio-scale process for ethanol production?

4. Consult companies dedicated to the production of wine and bioethanol in Colombia.
Bibliography.

- Jie Sun, Fei Wen, Tong Si, Jian-He Xu and Huimin Zhao, 2012
Direct Conversión of xylan to Ethanol by Minihemicellulosome Strains
Displaying an Engineered Recombinant Saccharomyces cerevisiae.
Applied and Enviromental Microbiology. 78 (11): 3837 - 3845.

- Jimenez, H.M. 2013. Industrial Biotechnology Laboratory Guidelines. Diploma in
Biotechnology (I). National Pedagogical University. Faculty of Science and Technology.
Department of Biology. Colombia.

Cybergraphy

- Wine   History: https://www.vinoseleccion.com/saber-de-vinos/historia-del-vino
Revision Date: 15/03/2019.

- Wyss, G & Bon van Elzakker, 2005. Grape production and wine making. Info
"Organic
HACCP" Available at: http://orgprints.Org/4928/l/14_YINO.pdf
                                        Revision date:
15/03/2019.

Extraction of nuclear DNA from *Saccharomyces cerevisiae*

By: Silvia R. Gómez D.

Introduction

The *Saccharomyces cerevisiae* yeast is unicellular, oval in shape, has no flagella, belongs to the Kingdom Fungi and is one of the main model organisms for understanding the cellular and molecular processes in eukaryotes. Currently, its impact goes beyond the production of food and beverages (bread, beer and wine), as it has been used as a food supplement to generate an increase in weight and height of birds, cattle and pigs and in the production of biofuel (Vasquez *et al* 2016).The DNA (deoxyribonucleic acid) is the hereditary material that is present in all living beings, containing the essential instructions for the development and functioning of all forms of life, it also stores and maintains the physical and functional genetic information that passes from generation to generation, allowing the maintenance of the species (Curtís, H., Chemically, it is a double-stranded molecule formed by the union of nucleotides (a nitrogenous base, a pentose, and a phosphate group through phosphodiester bonds (between a hydroxyl group (OH) in the 3' carbon of a nucleotide and a phosphate group (P04=) in the 5' carbon of the incoming nucleotide) being responsible for the DNA and RNA strand. The DNA strands are linked to form the chain through hydrogen bridges that bind the nitrogenous bases; Thiamine, Guanine, Adenine and Cytosine, See Figure 1. This model was proposed by Watson, Crick and Wilkins, in 1953 and is known as "the double helix model"; it was published in the journal Nature and describes what is known today as B-DNA. These researchers based their proposal on X-ray diffraction research reported by R. Franklin (Claros, 2003) The haploid genome of *S. cerevisiae* is small, compact with approximately 13 392 kb (excluding mitochondria, plasmids and RNA viruses) Mb and organized into 16 chromosomes in sizes ranging from 220 kb for chromosome 1 to 2352 kb for the XII (Madigan, M., et al 2010). At first glance, what is most striking about the genome is that 72% of it is genes, leaving very little room for non-coding DNA and other functional elements (Dujon, B. 1996). DNA extraction is one of the oldest techniques that have been performed in the field of molecular biology and dates back to 1869 when Miescher first isolated it from the pus of bandages used in hospitalized patients. After a simple treatment, he found that they were formed by a single chemical substance very homogeneous and not protein, which he called nucelein (substances rich in phosphorus located exclusively in the cell nucleus) (Claros,2003

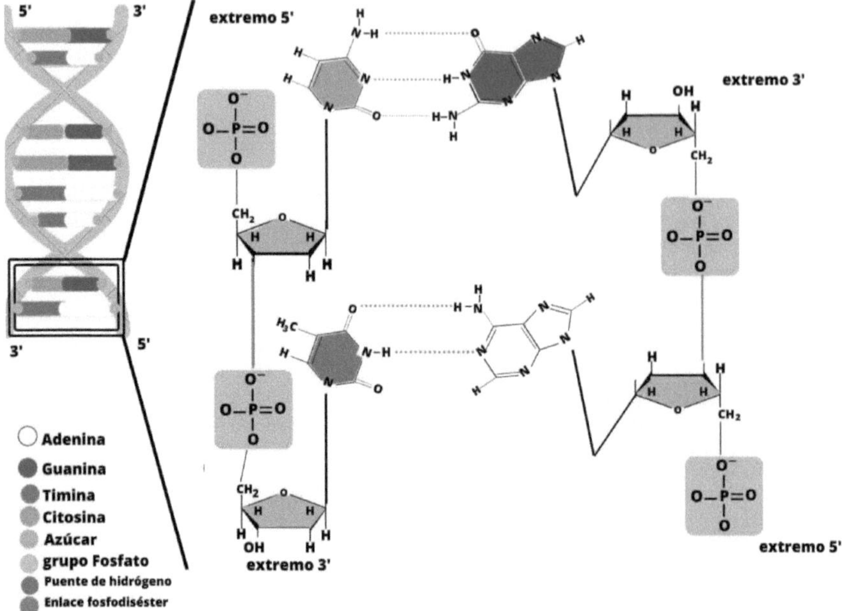

Figure **1.** Structure of DNA

To isolate DNA from the other cellular components such as proteins, carbohydrates, lipids and RNA, the same steps are generally used in all organisms with some variations in physical (maceration, temperature, centrifugation) and chemical (salt, detergent, enzymes, ethanol, EDTA, Tris etc.) methods depending on the degree of purity required for further procedures, amount of DNA required, amount of sample available and type of sample.

The steps are as follows:

1) Homogenization and/or concentration of the sample: with the first procedure, tissues are broken, which are inside a mortar and in a mechanical way using pistil and

elements such as sand or liquid nitrogen break down the cells; and with the second one through centrifugation the cells that are resuspended in liquid medium are concentrated.

2) Cellular lysis: consists of the rupture of the nuclear, cell membrane and/or cell wall without the degradation of the nucleic acids contained inside the cell. A lysis solution is used which is made up of EDTA (prevents DNA degradation because it traps the magnesium ions present), TRIS pH 8.0, (maintains the pH of the solution), salt (fragments the cell) and detergent such as CTAB, SDS, Triton X-100 or Sarkosly (breaks the lipid barrier by solubilizing the proteins and interrupting the lipid-lipid, lipid-protein and protein-protein interaction, due to its particular structure).

3)   Elimination of contaminants: It is the separation of nucleic acids from other cellular components (proteins, lipids and carbohydrates). High concentrations of salts, enzymes (meat tenderizer or k proteins) or organic solvents (phenol or chloroform: 24:1 isoamyl alcohol) can be used depending on the quality of DNA to be obtained.

4) DNA Precipitation: Isopropanol or 96% cold ethanol is used for this purpose, which allows the precipitation of DNA when it is in the presence of salts. It occurs due to the interaction between the positive charges of sodium that has been dissociated from the NaCl solution and the negative charges of DNA given by the phosphate groups. Subsequently, the salt residues are removed with 70% cold alcohol.

Objectives

- To perform nuclear DNA extraction from *Saccharomyces cerevisiae* yeast using

different protocols.

- Potentiate cognitive, social and verbal skills.

- Develop skills in the handling of laboratory instruments
- Understand the process of obtaining DNA taking into account the different physical and chemical methods used.

Materials, reagents and equipment for the first protocol

- Yeast: *Saccharomyces cerevisiae* Microcentrifuge
- Incubator
    - Eppendorf tube rack.
    - 1.5 mL Eppendorf tubes
    - Paper towels.
    - 1000 pL micropipettes
    - Blue tips for Micropipettes
    - Sterile water
    - 70% cold ethanol
    - Isopropanol 100% cold
    - Fine tip marker for tubes
    - Nitrile gloves
    - Protective eyewear

- Lab coat
- Matches
- Ice
- Potassium acetate (KAc) 5M (pH 4.8)
- YPED liquid culture medium (yeast extract 1%, peptone 2%, glucose 2%).
- R
ouar
y
bath
-
Micr
osco
pe
- Slide and cover sheets
- Cell wall lysis solution (1M sorbitol solution, 0.1M EDTA, (pH 5), with 50
Units (U) of the enzyme Beta-glucoronidase (Sigma-Aldrich)
- Solution with Tris-HCl 50 mM, EDTA 20 mM (pH 7.4)
- SDS at 10%.
- Shaking plate
- TE (Tris-EDTA pH 7.4)

Materials and reagents for the second protocol (homemade)

- 1 beaker of 500 my
- 2 x 50 m falcon tube with cover
- 1 mortar with its pistil
- 1 heating plate or stove
- A tube clamp
- 5 grams of dry yeast.
- Lysis solution (10 mi of H2O and 1 gr of detergent bar)
- 0.8 gr of common salt
- 0,5 gr of meat tenderizer
- 30 mi of ethanol or 96% alcohol
- Water for the beaker.

Methodology

First protocol: Taken from Osorio-Cadavid, Esteban; Ramírez, Mauricio; López William Andrés and Mambuscay Luz Adriana (2009)

1. Let the yeast cultures grow for 16 hours in agitation (120 rpm) or until more than 100 million cells/ml are obtained at 28°C, in 5 mi of YPED (yeast extract 1%, peptone 2%, glucose 2%).

2. Collect cells by centrifugation in an Eppendorf tube at 8000 rpm for 5 min.

3. Discard the supernatant and resuspend in 0.5 mi a solution of 1M sorbitol, 0.1M EDTA, (pH 5), containing 50 Units (U) of the enzyme Beta-glucoronidase (Sigma- Aldrich).

4. Then incubate in a water bath at 37 °C for 60 min and shake periodically (sometimes monitor the loss of the cell wall by placing the cells in distilled

water and checking the reduction of cell density under the optical microscope).

5. Centrifuge the spheroplasts at 8000 rpm for 10 min and resuspend the precipitate in 0.5 mi of Tris-HCl 50 mM, EDTA 20 mM (pH 7.4).

6. Then add 50pl of 10% SDS and incubate in a water bath at 65°C for 30 minutes.

7. After this time add 0.2 mi of potassium acetate (KAc) 5M (pH 4.8), and resuspend it for at least 30 sec. Then leave it on ice for 30 min.

8. Then centrifuge at 14000 rpm for 5 min and transfer the supernatant to another tube.

9. Centrifuge again for 5 min to remove other impurities and transfer the supernatant to another tube.

10. Add 1 mL of 100% cold isopropanol and incubate at room temperature for 5 min while gently shaking the tube.

11. Centrifuge at 14000 rpm for 10 min and discard the supernatant.
12. Add 0.5 mL of 70% cold ethanol and centrifuge at 14000 rpm for 5 min.

13. Discard the supernatant and let the precipitate dry at room temperature.

14. Finally, resuspend the DNA in 50 pl of TE (Tris-EDTA pH 7.4) or distilled water. The samples are kept at -20 °C until further use.

Second protocol: home procedure

1. Take a beaker with 300 mi of water and place it on the stove or heating plate and let it boil.

2. Meanwhile, macerate by homogenizing well in a mortar 5 gr of yeast with the lysis solution (10 mL distilled water and 1 gr of detergent).

3. Add the contents of a 50 mL falcon tube and place it in the beaker containing the boiling water (bain-marie) for 30 minutes. Mix the contents of the tube every 2 min by inversion and use the tweezers to hold it.

4. After the time has elapsed, remove the tube and let it cool down.

5. Then add 0.5 gr of meat tenderizer and mix gently by immersion.

6. Then add 0.8 gr of salt and mix gently by immersion.

7. Gently transfer the entire contents to a new 50-meter falcon tube.

8. Finally, slowly add three volumes (three times the volume of sample found in the tube) of ethanol or 96% cold alcohol to the walls of the tube and wait for the DNA to become evident.

Questionnaire

1. Explain why the detergent produces lysis of the cells?

2. What effect does the enzyme Beta-Glucoronidase have on the DNA extraction procedure?

3. What is the purpose of salt in DNA extraction?

4. How does the meat tenderizer act in the cell?

5. What is the use of 100% and 70% cold alcohol?

Bibliography

- Madigan, M., Martinko, J., Parker, Brock J. (2010) Brock, Biology of Microorganisms. 10th edition. Prentice Hall.

- Clear, Gonzalo. (2003). Historical approach to molecular biology through its protagonists, concepts and fundamental terminology. *Panace@*, 4(12), 168-179. Retrieved from http://www.medtrad.org/pana.htm.

- Curtís, H. and Barnes, N. Sue. 2000. Biology. Panamerican Medical Ed. 6th Edition. Madrid Spain.

- Dujon, B. (1996). The yeast genome project: what did we learn? *Trends Genet.* 12: 263-270.

- Osorio, E.; Ramírez, M; López W. y Mambuscay, L (2009) Standardization of a simple protocol for the extraction of genomic DNA from yeasts Revista Colombiana de Biotecnología, vol. XI, núm. 1, julio, 2009, pp. 125-131 Universidad Nacional de Colombia Bogotá, Colombia

- Vásquez J. A., Ramírez C. M. and Monsalve Z. I. 2016. Update on molecular characterization of yeast 129 Rev. Biotechnology. Vol. XVIII No. 2 129-139

In vitro **cellulase screening from *Penicillium aurantiogriseum*.**

By: Hugo Mauricio Jimenez M.

Introduction

The fungi have a great potential for the production of diverse chemical compounds of importance for the humanity, either from a medical, agronomic, industrial or environmental perspective, the use of fungi and/or their products, are associated with technological processes of application to the Industry or the Environmental Management (Jiménez, H., 2020).

An example of the above is the production of enzymes, which are biological catalysts of high efficiency and specificity to the substrate, achieve a reaction speed of 10^3 and 10^7 times faster than non-catalyzed reactions, as is the case of cellulases, which are glycosyl hydrolases, and use two mechanisms of hydrolysis of the glycosidic bond of cellulose found mainly in wood, plant debris such as leaf litter and tree branches.

The woods are composed mainly of three structural polymers, lignin, cellulose and hemicellulose. Lignin is resistant to physical and chemical degradation, as are the cellulose fibers that are embedded in a matrix of hemicellulose and lignin, therefore their degradation is difficult. *Phylum Basidiomycota, Ascomycota, Glomeromycota and Neocallimastigomycota* fungi produce a high number and variety of extracellular enzymes for the degradation of wood and plant debris with high lignocellulose content.

Cellulose is the most abundant component that exists on Earth, it is produced by plants forming part of their cell wall, cellulose is one of the most used materials since ancient times, and nowadays it is the source of organic fuels, chemical compounds, fibers and materials needed to cover human needs such as paper, pulp, wood, etc. It is estimated that around 180 billion tons of cellulose are produced by plants annually, making it one of the most important renewable carbon sources on Earth (Martinez, C., et al, 2008).

Cellulose is a linear polymer composed of glucose residues joined by B bonds (1-4), through which they bind to molecules of cellobiose, disaccharide composed of two glucose residues.

Due to the recalcitrant nature of cellulose, certain organisms such as bacteria and fungi produce the necessary enzymes to use it (Béguin & Aubert, 1994). Two important groups with cellulolytic capabilities have been identified. The first is the anaerobic group,

29

which comprises bacterial and fungal species inhabiting sewage and the rumen and intestinal tract of herbivorous animals and some insects such as beetles and termites (Cazemier et al., 2003; Warnecke et al., 2007). Bacterial examples belonging to this group, among others, are the genera *Clostridium* and *Ruminococcus*. While some identified fungi are: *Anaeromyces mucronatus, Caecomyces communis, Cyllamyces aberencis, Neocallimcistix frontalis, Orpinomyces sp. and Piromyces* sp. (Doi, 2007; Teunissen & Op den Camp, 1993). The second group includes aerobic soil-dwelling species, especially in forests, such as *Cellulomonas* (Elberson et al., 2000) and *Streptomyces* (Alani et al., 2008), and basidiomyces wood rot fungi (Baldrian & Valaskova, 2008; Martinez et al., 2005). White wood rot is carried out in 96% of the cases by fungi of the *Polyporaceae* family, such as *Panus, Polyporus, Pycnoporus and*

Trametes (Martínez et al., 2005), references cited by Martínez, C., et al, 2008.

The concept of using cellulose as a raw material for sugars that can be bioconverted into fuels through the use of microorganisms has regained great importance in recent years due to high oil prices (Sun & Cheng, 2002, Quoted by Martinez, C., et al, 2008).

The main challenge of the industry and biotechnology in the production of biofuels such as bioethanol is the bioconversion of cellulose, so cellulases have acquired great importance in these processes (Martinez, C., et al, 2008).

The enzymatic action for the hydrolysis of cellulose involves the sequential operation and synergy of a group of cellulases, which have different binding sites, due to the complex nature of the cellulose molecule. The typical cellulase system includes three types of enzymes: endo-P-l,4-glucanase (Cx) (1,4-P-D-glucan glucanohidrolase E.C.3.2.1.4), exo-P-1,4-glucanase (Cl) (1,4-P-D-glucan cello-biohydrolase E.C.3.2.1.91) and P-l,4-glucosidase (cellobiase) (Cb) (P-D-glucoside glucohydrolase E.C.3.2.1.21) (Hahn-Hágerdal & Palmqvist 2000, Cited by Chacón, O. & k, Waliszewski, 2005).

Agricultural residues are rich in cellulose, hemicellulose and lignin; they can be used as substrates for the cultivation of filamentous fungi capable of producing extracellular enzymes with cellulase activities, with important industrial applications, such as the hydrolysis of lignocellulosic biomass for the production of ethanol (Rodríguez, I. and Pifieros, Y. 2007)

Bio-alcohol is a fuel of vegetable origin that has the characteristics similar to those of fossil fuels, which allows its use in slightly modified engines. Furthermore, biofuels do not contain sulphur, one of the causes of acid rain. Bioethanol can be manufactured from any organic raw material that contains significant amounts of sugars (Ballesteros, M. 2006, cited by Paredes Medina et al, 2010).
The cellulases are used to degrade cellulite compounds such as bagasse and obtain fermentable sugars such as glucose and thus obtain bioethanol, these technologies are already being applied in Colombia.

Objectives

- To carry out *in vitro* bioassays for the detection of cellulase production by the microfungus *Penicillium aurantiogriseum.*

- Acquire skills in assembling *in vitro* bioassays.

- Develop skills in the handling of laboratory instruments

- Potentiate scientific cognitive skills.

Materials, reagents and equipment

- Microfungus *Penicillium aurantiogriseum.*
- Erlenmeyer
- Distilled water.

- Potato Dextrose Agar Culture Media (PDA).
- Culture medium for amylolytics.
-Aluminum paper.
- Paper.
- Masking tape.
- Autoclave.
- Laminar Flow Chamber.
- Petri dishes.
- Handle for microfungi.
- Lugol
- Incubator.

Methodology

Preparation of Potato Dextrose Agar Culture Media (PDA)

The volume of the PDA Culture Media for each petri dish is 25 mL, so it is necessary to make the respective calculations from the Oxoid Potato Dextrose Agar flask, weigh the

31

quantity, add it to an Erlenmeyer with the required volume of distilled water, cover with aluminum foil and reinforce with paper and masking tape, sterilize in the autoclave together with the petri dishes, after sterilization serve the culture medium in the petri dishes in the laminar flow chamber.

Preparation of Culture Media for Cellulolytics (MCC)

The volume of Cellulolytic Culture Media (CCM) for each petri dish is 25 mL, so it is necessary to make the respective calculations from the g/L composition of CCM: Microcrystalline Cellulose (Merck) 5 g, NH_4NO_3 1 g, Saline Solution (NaCl 5%) 10 mL, Agar - Agar 20 g, and Distilled Water 1000 mL, weigh the amount, add it to an Erlenmeyer with the required volume of distilled water, cover with aluminum foil and reinforce with paper and masking tape, sterilize in the autoclave together with the petri dishes, once the sterilization is over serve the culture medium in the petri dishes in the laminar flow chamber.

Activation of *Penicillium aurantiogriseum*

From a strain of *Penicillium aurantiogriseum,* sow a fragment of the microfungus in the center of a Petri dish with the PDA culture medium and leave it at room temperature for 7 days.

Cellulase screening bioassays.

From the PDA Petri dish with *Penicillium aurantiogriseum* sow a fragment of it in the center of a Petri dish with the Cellulolytic Culture Medium (CCM) and leave it at room temperature for 7 days, then add Congo Red on the culture, after 15 minutes observe and measure with a ruler the yellow halo around the sowing.

Questionnaire.

1. Why do you see a yellow halo around the planting of the microfungus when you add Congo Red?

2. Why do you see a reddish coloration when adding Congo Red to the cellulolytic culture medium (CMC)?

3. See other tests for enzymes.

4. See other microorganisms used for cellulase production.
5. How would a Bioprocess of scaling for the production of cellulases by *Penicillium aurantiogriseuml*

Bibliography.

- Chacon, O. & k, Waliszewski, 2005. Commercial Cellulase Preparations and Applications in Extractive Processes. University and Science. Humid Tropics. 21 (42) pp 111-120. Available at: http://era.ujat.mx/index.php/rera/article/viewFile/337/273. Review Date: 03/26/2019.

Reference cited in this article:

> Hahn-Hagerdal B, Palmqvist E (2000) Fermentation of lignocellulosic hydrolysates. II: Inhibitors and mechanismsof inhibition. Bioresource Technology. 74: 25-33.

- Couturier M, et al. 2011. *Podospora anserina* hemicellulases potentiate the *Trichoderma reesei* secretóme for saccharification of lignocellulosic biomass. Appl. Environ. Microbiol. 77: 237 - 246.

- Jimenez, H.M., 2020. Syllabus Mushroom Biology Seminar. National Pedagogical University. Department of Biology. Colombia.

- J-G, Berrín, Navarro, D., Couturier, M. and L, Meessen, 2012. Exploring the Natural Fungal Biodiversity of Tropical and Températe Forests toward Improvement of Biomass Conversión. Appl. Environ. Microbiol. 78 (18): 6483 - 6490.

- Martínez, C, Balcázar, E., Dantán, E and J L.Folch-Mallo. 2008. Fungal cellulases: Biological aspects and applications in the energy industry. Latin American Journal of Microbiology. Vol. 50. No. 3 and 4. Pp 119 -131. Available at http://www.medigraphic.com/pdfs/lamicro/mi-2008/mi08-3_4i.pdf Revision date: 26/03/2019.

References cited in this article:

> - Alani, F., Anderson, W. & Moo-Young, M. 2008. New isolateof Streptomyces sp. with novel thermoalkalotolerant cellulases.Biotechnol Lett. 30, 123-126.

Béguin, P. & Aubert, J.-P. 1994. The biological degradation ofcellulose. FEMS Microbiol Rev. 13, 25-58

Baldrian, P. & Valaskova, V. 2008. Degradation of celluloseby basidiomycetous fungí. FEMS Microbiol Rev. Epub aheadof print, doi: 10.111 l/j.1574- 6976.2008.00106.x.

Cazemier, A. E., Yerdoes, J. C., Reubsaet, F. A., Hackstein, J.H., van der Drift, C. & Op den Camp, H. J. 2003. Promi-cromonospora pachnodae sp. nov., a member of the(hemi)cellulolytic hindgut flora of larvae of the scarab beetlePachnoda marginata. Antonie Van Leeuwenhoek. 83, 135-48.

Doi, R. H. 2007. Cellulases of mesophilic microorganisms: cel-lulosome & non- cellulosome producers. Doi 0:14190021

Martínez, A. T., Speranza, M., Ruiz-Dueñas, F. J., Ferreira, P.,Camarero, S., Guillén, F., Martínez, M. J., Gutiérrez, A. & delRío, J. C. 2005. Biodegradation of lignocellulosics: microbial,Chemical, and enzymatic aspects of the fungal attack oflignin.Int Microbiol. 8, 195-204.

Sun, Y. & Cheng, J. 2002. Hydrolysis of lignocellulosicmaterials for ethanol production: a review. Bioresour Tech-nol. 83, 1-11 Teunissen, M. J. & Op den Camp, H. J. 1993. Anaerobic fun-gi and their cellulolytic and xylanolytic enzymes. AntonieVan Leeuwenhoek. 63, 63-76. Wamecke, F., Luginbuhl, P., Ivanova, N., Ghassemian, M.,Richardson, T. H., Stege, J. T., Cayouette, M., McHardy, A.C., Djordjevic, G., Aboushadi, N. et al. 2007. Metagenomicand functional analysis of hindgut microbiota of a wood- feeding higher termite. Nature. 450, 560-5.

- Paredes Medina, Álvarez Núfiez, and M. Silva Ordofies. 2010. Obtaining Cellulase Enzymes by Solid Fermentation of Mushrooms to be Used in the Process of Obtaining Bioalcohol from Banana Farming Residues. Technological Magazine ESPOL - RTE, Yol. 23, N. 1, 81-88.

Reference cited in this article:

Ballesteros, M. 2006, "Carburantes sin petróleo: Bioetanol", Investigación y ciencia, ISSN 0210-136X, No. 362, 2006. pp. 78-85.

- Rodríguez, I. and Piñeros, Y. 2007. "Production of cellulolytic enzymatic complexes by means of the cultivation in solid phase of *Trichoderma sp.* on the empty bunches of oil palm as substrate", Group of Exploitation of Agrifood Resources, Food Engineering Program, Jorge Tadeo Lozano University, Bogotá, Colombia.

In vitro amylase production tests with *Aspergillus fumigatus*

By: Hugo Mauricio Jimenez M.

Introduction

Starch is a semi-crystalline glucose polymer that is abundant in nature and is obtained mainly from corn, wheat, rice, and potatoes. If it comes from a tuber it is usually called starch (potato starch); if it comes from a cereal, starch. The properties of starch vary depending on the product from which it is extracted and the variety (Castells, P. 2009).

Starch is a digestible complex carbohydrate (polysaccharide) from the group of glucans. It consists of glucose chains with linear (amylose) or branched structure (amylopectin). It constitutes the energy reserve of plants (Castells, P. 2009).

Amylose is a glucose polymer that contains 1,000-4,000 units of this monomer, and therefore has a molecular weight of 200,000-800,000 daltons, a value that varies not only according to the species of plant, but also within the same species and depends on the state of maturation (Hoseney, 1991, cited by Espitia, L., 2009).

Each unit of glucose is linked to the next by a glycidic a-1.4 bond, this bond determines that the glucose reducing group is located in position 1. The long linear nature gives amylase some unique properties, such as its ability to form complexes with iodines, alcohols or organic acids. At room temperature, the chain of glucose molecules adopts a spiral conformation whose helixes allow to lodge in its interior a molecule of iodine. When amylose is treated with iodine, the iodine is located in those forming an amylo-iodine complex that has a blackish blue color (Hoseney, 1991, quoted by Espitia, L., 2009).

Amylopectin is a polysaccharide whose main chains are glucose traces linked to 1-4, as in amylose, and sporadically have branches at 1-6 located every 15-25 linear units of glucose. Their molecular weight is very high (Bailey and Bailey, 1998, cited by Espitia, L., 2009).

Starch is degraded by the action of two enzymes, a - amylase and B - amylase.

The enzyme a - amylase catalyzes the random hydrolysis of the a-1,4 glycoside bonds of the central region of the amylose and amylopectin chain, except for molecules close to the branching, obtaining maltose and oligosaccharides of various sizes (Crueger and Crueger, 1993, cited by Espitia, L., 2009).
The *B - amylase or a - 1.4 glucan - maltohydrolases enzyme is an exoenzyme that attacks the a - 1.4 glycoside bonds on the outside of the starch chain, the 13 - amylase separates maltose units from the non-reducing ends of this alternate hydrolysis of glycoside bonds (Pedroza, 1999, cited by Espitia, L., 2009).

Amylases are used in the manufacture of bread by degrading the starch in the flours into fermentable sugars for yeast activation.

Some amylases are used as detergents to dissolve starches in certain industrial processes, such as the production of pasta.

Amylase enzymes, due to their hydrolytic capacity, have been very significant in the last decades in several industries that have seen a better way to optimize their processes by applying enzyme-based biotechnology techniques (Durango E., 2008).

Industries such as bakery, pastry, food, textile, beer, paper, sugar among many others have seen in these proteins a great opportunity for trade. Amylases occupy about 25% of the enzyme market, completely replacing chemical hydrolysis processes in the starch industry. Due to the thermostability of this enzyme, industrially amylases have great applicability in various processes (Durango E., 2008).

From starch and use of amylases can be obtained syrups of different composition and physical properties syrups are used in a variety of foods such as soft drinks, candy, baked goods, ice cream, sauces, baby food, canned fruits and preserves (Durango E., 2008).

Of all the applications on an industrial scale, the most efficient is the production of high fructose corn syrup (HFCS), the purpose of this process is to obtain a material with a sweetening power similar to sucrose from a low price raw material such as corn starch (Durango E., 2008).

The producing sources of a-amylases include plants, animals and microorganisms, and it is the microbial enzymes that are most in demand in industrial applications (Grupta. R., et al, 2003, cited by Espinel, E and E. López, 2009).

Traditionally the production of a -amylases has been carried out by submerged liquid fermentation (FLS) processes due to the greater control of environmental factors such as temperature and pH, however, solid phase fermentation (FFS) constitutes an interesting alternative since the metabolites are concentrated, and the purification processes are less expensive (Pandey, A. et al, 2000., Soni, S., 2003, cited by Espinel, E and E. López, 2009).

Objectives

- To carry out *in vitro* bioassays for the detection of amylase production by the microfungus *Aspergillus fumigatus*.

- Acquire skills in assembling *in vitro* bioassays.

- Develop skills in the handling of laboratory instruments

- Potentiate cognitive and scientific skills.

Materials, reagents and equipment

- *Microfungus Aspergillus fumigatus.*
- Erlenmeyer
- Distilled water.
- Potato Dextrose Agar Culture Media (PDA).
- Culture medium for amylolytics.
-Aluminum paper.
- Paper.
- Masking tape.

- Autoclave.
- Laminar Flow Chamber.
- Petri dishes.
- Handle for microfungi.
- Lugol
- Incubator.

Methodology

Preparation of Potato Dextrose Agar Culture Media (PDA)

The volume of PDA Culture Media for each petri dish is 25 mL, so it is necessary to make the respective calculations from the Oxoid Potato Dextrose Agar flask, weigh the quantity, add it to an Erlenmeyer with the required volume of distilled water, cover with aluminum foil and reinforce with paper and masking tape, sterilize in the autoclave together with
the petri dishes, after sterilization serve the culture medium in the petri dishes in the laminar flow chamber.

Preparation of Culture Media for Amylolytics (MCA)

The volume of Culture Media for amylolytics (MCA) for each petri dish is 25 mL, so it is necessary to make the respective calculations from the g/L composition of MCA: Soluble starch 10 g, Na2HP04 3 g, MgSCfl 7H2O 0.1 g, Agar - Agar 20 g, and Distilled Water 1000 mL, weigh the amount, add it to an Erlenmeyer with the required volume of distilled water, cover with aluminum foil and reinforce with paper and masking tape, sterilize in the autoclave together with the petri dishes, once the sterilization is over serve the culture medium in the petri dishes in the laminar flow chamber.

Activation of *Aspergillus fumigatus*

From a strain of *Aspergillus fumigatus,* sow a fragment of the microfungus in the center of a Petri dish with the PDA culture medium and leave it at room temperature for 7 days.

Amylase screening bioassays

From the PDA Petri dish with *Aspergillus fumigatus,* sow a fragment of it in the center of a Petri dish with the culture medium for amylolytics (MCA) and leave it at room temperature for 7 days, then add Lugol on the crop, after 10 minutes observe and measure with a ruler the yellow halo around the sowing.

Questionnaire.

1. Why do you see a yellow halo around the planting of the microfungus when adding Lugol?

2. Why is a purple coloration observed when Lugol is added to the amylolytic culture medium (ACM)?

3. See tests for detection and quantification of amylases.

4. See other microorganisms used for the production of amylases.
Bibliography

- Castells, P. 2009. The Starch. Research and Science. No. 396.

- Espinel, E., and E. López, 2009. Purification and characterization of a-amylase from *Penicillium commune* produced by solid phase fermentation. Revista Colombiana de Química, vol. 38, no. 2, pp. 191-208 Universidad Nacional de Colombia. Bogotá.

References cited in this article:

> Gupta, R.; Gigras, H.; Mohapatra,H.; Goswami, V.; Chauhan, B. Microbial a-amylases: a biotechnological perspective. Process Biochemistry. 2003: 1599-1616.3.

> Pandey, A.; Soccol, C.R. New developments in solid State fermentation: Ibioprocesses and products. ProcessBiochemistry.2000:1153-1169

> Soni, S. K.; Arshdeep, K.; Gupta, J.K. A solid State fermentation based bacterial a-amylase and fungal glucoamylase system and its suitability for the hydrolysis of wheat starch. Process Biochemistry. 2003:185-192.

- Espitia, L. 2009. Determination of the concentration of commercial Alpha and Beta amylases in the production of ethanol from barley using *Saccharomyces cerevisiae*. Grade work. Industrial Microbiology Department. Science Faculty. Pontifical Javeriana University.

References cited in this article:

> Bailey, P.S. and C.A., Bailey. 1998. Organic Chemistry: Concepts and Applications. 5th Edition. Prentice Hall Edition. Mexico.

> Crueger W and A Crueger, 1993. Biotechnology: Manual of Industrial Microbiology. Acribia Publishing.

> Hoseney, R., 1991. Principles of Cereal Science and Technology. Acribia, ed. Zaragoza, Spain.
> - Pedroza, 1999. Production of *thermostable* amylase from *Thermus* sp. Master's thesis. Industrial Microbiology Department. Science Faculty. Pontifical Javeriana University.

Crude Oil Biodegradation Bioassays with

Pseudomonas fluorescens.

By: Hugo Mauricio Jimenez M.

Introduction

Pseudomonas fluorescens was discovered by Migula in 1895, they are aerobic chemorganotrophic bacteria, their metabolism is based on oxide-reduction reactions to obtain energy. It is a straight gram (-) bacillus from 0.5 - 0.8 pm. It presents loophotropic polar flagella. Its optimum temperature is 25 °C to 30 °C, although there are reports of 5 °C and 42°C. It is mainly found in the rhizosphere, it can also be found in soil as saprophyte and in water (Boresi, M. 2009).

It solubilizes phosphates in two ways: first by the production of organic acids such as citric acid and oxalic acid that act on the pH of the soil which solubilizes the inorganic phosphorus and releases the phosphates into the soil, the other way is by the production of phosphatases that act on the ester bonds releasing the phosphate groups from the organic matter of the soil (Boresi, M. 2009).

They produce plant growth stimulating hormones, such as auxins, gibberellins and cytokinins, and also stimulate seed germination.

It degrades contaminants such as styrene, TNT and polycyclic aromatic hydrocarbons (López, J. et al. 2006).

P. fluorescens uses various petroleum substrates such as total hydrocarbons (TPH) and polychlorinated biphenyls (PCB's), aerobically biodegradable aromatic hydrocarbons such as naphthalene and phenanthrene (López, J. et al. 2006).

Improper management of hazardous waste has generated a worldwide problem of soil, air and water pollution. Among the most serious contaminations is the extraction and handling of crude oil in the producing countries (López, J. et al. 2006).

In Colombia, the transportation of crude oil and its derivatives has been significantly affected over the past 30 years by ongoing terrorist activity against oil pipelines and facilities that has generated countless blasts and oil spills (López, J. et al. 2006).
Oil contamination of environments is considered to be of high persistence and affects the balance of ecosystems. The fragility of these is such that nature does not have the facility to biodegrade oil in an easy and fast way. One liter of crude oil occupies an area of approximately half a soccer field in the aquatic environment (Lozano, N. 2005).

The decomposition of oil by the microbial route is an agile and safe mechanism to eliminate contamination. That is why it is important to study the ways in which microorganisms assimilate oil compounds and how the decontamination process can be accelerated. Bioremediation technologies have been developed in which microorganisms or plants act that allow the decomposition of toxic compounds (Lozano, N. 2005).

Bioremediation is a "friendly" alternative to the progressive deterioration of the quality of the environment due to the constant spill of crude oil that contaminates the soil, air, and water, since this problem affects public health, as well as the extinction of flora and fauna in Colombian ecosystems (López, J. et al. 2006).

Crude oil is composed of hydrocarbon substances and small amounts of sulfur, nitrogen and oxygen, which make it difficult to biodegrade. Petroleum hydrocarbons have from one to 50 or more carbon atoms and have a variety of molecular forms such as kerosenes, naphthalenes and aromatics (Lozano, N. 2005).

The biodegradation of crude oil by the microbial route is a fast and safe mechanism to eliminate contamination, a bacterium used for this purpose is *Pseudomonas fluorescens* that uses various oil substrates such as Total Hydrocarbons (TPH) and Polychlorinated Biphenyl (PCB's) as a source of carbon and energy and aerobically biodegrades aromatic hydrocarbons such as naphthalene and phenanthrene, some of these hydrocarbons such as methane, are made up of few atoms and are gaseous at room temperature; others, such as the dean, are heavier and less volatile. At low temperatures some are gaseous, such as propane, while others are solid, such as kerosene and asphalts (Lozano, N. 2005).

Bioremediation practices consist of the use of microorganisms such as bacteria and microfungi, and plants to neutralize toxic substances, transforming them into less toxic or non-toxic substances for the environment and human health.

Before conducting bioremediation programs for soils or water bodies contaminated with crude oil, it is important to first conduct crude oil biodegradation bioassays at
The study of the physiology, culture conditions and growth of the microorganisms to be used.

Objectives

- Perform a small-scale Crude Oil Biodegradation Bioassay using the bacterium *Pseudomonas fluoride escens*

- Observing Crude Oil Biodegradation from *Pseudomonas fluorescens* bacteria.

- Develop skills and abilities in the handling of laboratory equipment and instruments.

- Potentiate scientific skills.

Materials, Equipment and Reagents

- Nitrile gloves
- Lab coat
- Selective Culture Medium Pseudomonas Agar Base (Petri dishes and inclined test tubes).

- Pure culture of *Pseudomonas fluorescens*
- Alcohol lighter.
- Bacteriological round handle
- Sterile distilled water.
- 500 mL bottles.
- Minimum Medium Salt (MMS).
- Crude oil.

Methodology

Preparation of Culture Media for *Pseudomonas fluorescens* activation

The volume of Culture Media for each petri dish is 25 mL, so it is necessary to make the respective calculations from the Oxoid Bottle of Pseudomonas Agar Base, weigh the quantity, add it to an Erlenmeyer with the required volume of distilled water, cover with aluminum foil and reinforce with paper and masking tape, sterilize in the autoclave together with the petri dishes, once the sterilization is over serve the culture media in the petri dishes in the laminar flow chamber.

From a strain of *Pseudomonas fluorescens* sow it by isolation - exhaustion technique in Petri dishes with the Selective Culture Medium Pseudomonas Agar Base (MCSPF) and leave it in incubation at 30 °C for 48 hours.

Preparation of *Pseudomonas fluorescens* inoculum

From the Petri dishes with the sowing by exhaustion of *Pseudomonas fluorescens* in the culture medium (MCSPF) sow it again (place 5 roasts) in a 100 mL Erlenmeyer with 30 mL of nutritive broth and leave it in incubation at 30 °C for 48 hours (this is the inoculation of
Pseudomonas fluorescens).

Crude Oil Biodegradation Bioassays.

In 4 bottles of 500 mL, place in each one 285 mL of the Minimum Salt Medium (MMS) sterile composition g/L: KH_2PO_4 5 g, NH_4Cl 10 g, Na_2SO_4 20 g, $KNOs$ 20 g, $CaCl_2$ $6H_2O$ 0.01 g, $MgSO_4$ 1g, $FeSO_4$ 0.004g, Distilled Water: 1000 mL.

Add 5 mL of *Pseudomonas fluorescens* inoculum to each of the 3 vials. To the remaining vial, leave it without the inoculum of the bacteria, this would be the negative control. To each vial add 15 mL of crude oil. Cover each vial with sterile gauze.

Leave at room temperature and observe thinning of the crude oil layer every 8 days, always compare with the negative control.

Questionnaire

1. See the scientific name of 5 other bacteria used in Crude Oil Biodegradation.

2. Name and describe 3 marine tanker accidents in which crude oil spills occurred.

3. In Colombia, explain the 2 reasons why there are crude oil spills in soil and water.

4. Explain the difference between biodegradation and crude oil bioremediation.

Bibliography

Boresi, M. 2009. *Pseudomonas fluorescens.* Microbiology - Missouri. http://web.mst.edu/microbio/BI0221 -2009/P_fluorescens html

- MPLM, 2016. Industrial Microbiology Laboratory Practice Manual: Bioassays of Biodegradation of Crude Oil with Bacteria University of the Andes. Microbiology Department. Colombia.

- Lozano, N. 2005. Bioremediation of oil-contaminated environments. Tecnogestion, a look at the environment. Vol. II, No. 1. Pages 51-55.

- López, J. et al. 2006. Bioremediation of Soils Contaminated with Petroleum-Derived Hydrocarbons. NOVA. Vol. 4. No. 5. Pages 82 - 90.

By: Silvia R. Gómez D.

Introduction

Plasmids are small circular or linear DNA molecules capable of replication independently of the central chromosome of the host cell, are stable, are in the cytoplasm, occur in large numbers of copies (polyplasmia) and by a process called conjugation are transmitted between cells, See Figure 2 (Madigan, M., *etal2010*).

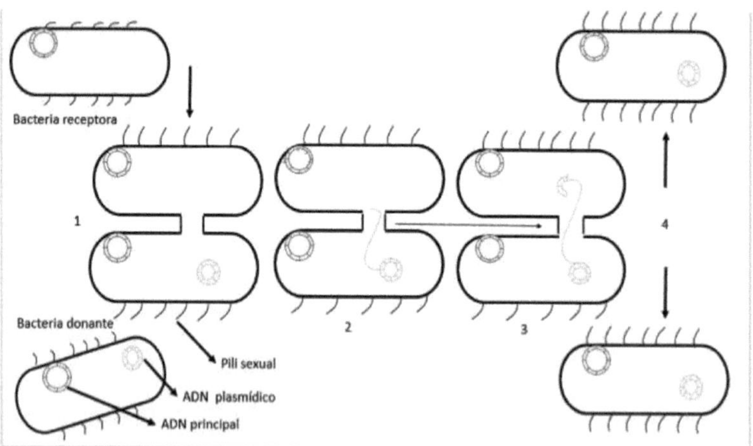

They are generally found in bacteria, although they are also usually found in yeasts and not
1: Formación de puente de conjugación. 2 transferencia de una hebra de ADN plasmídico.
3 Síntesis de la hebra complementaria 4.Las bacterias se separan conteniendo el ADN plasmídico

Figure 2. Conjugation process

They are generally found in bacteria although they are also usually found in yeasts and are not essential for the cell, they contain gene or genes that provide them with selective or adaptive advantages when found in an organism. Table 1 lists different characteristics provided by plasmids and examples of bacteria.

Chemically, plasmid DNA is a double helix with a right turn (dextrogenic) that has a phosphate and sugar skeleton on the outside and nitrogenous bases on the inside joined by hydrogen bridges (Curtís, H., Bames, N. S. *et al.* 2007).

Plasmids are widely used in molecular biology as vectors for the introduction of genes, which after being cloned into them can be incorporated into different organisms. At the laboratory level, plasmids are easy to isolate, introduce and manipulate into a host cell due to their small size. A phenomenon of considerable importance both in plasmid research and in evolution and ecology is incompatibility. When a plasmid is inserted into a cell containing another plasmid, the latter often cannot be maintained by being lost during the cell replication process. For this reason they are said to be incompatible. The incompatibility is controlled by genes found in them and there are many incompatible groups (Madigan, M., *et al* 2003).

Table 1. Phenotypes obtained by plasmids in bacteria Taken Madigan, M., et al 2010

Phenotype class	Organism
Production of antibiotics	*Streptomyces*
Conjugation	*Pseudomonas*
Physiological functions	
Octane degradation, camphor	*Pseudomonas*
Herbicide degradation	*Alcaligenes*
Acetone and butanol formation	*Clostridium*
Use of lactose, urea and nitrogen fixation	Enteric bacteria
Symbiotic nitrogen nodulation and fixation	*Rhizobium*
Pigment production	*Staphylococcus*
Resistance	

Antibiotic resistance	*Staphylococcus*
Resistance to cadmium, cobalt, mercury, nickel and/or zinc	*Pseudomonas*
Resistance to bacteriocines (and production)	*Bacillus*
Virulence	
Host cell invasion	*Salmonella*
Coagulase, hemolysin, enterotoxins	*Bacillus, Staphylococcus*
Enterotoxins and K-antigen	*Escherichia*
Tumorogenicity in plants	*Agrobacterium*

To isolate plasmid DNA from proteins, carbohydrates, lipids and RNA, the same steps are generally used in all living beings; these are: a) sample homogenization or sample concentration b) cell lysis, c) removal of contaminating molecules and d) DNA precipitation, with some variations in the physical and chemical methods depending on the degree of purity required.

The most widely used method of plasmid DNA extraction is alkaline lysis, which takes advantage of the differences in size and degree of twisting of the plasmid DNA with respect to the bacterium's primary DNA. The bacterial master DNA is a single, circular, large molecule that is not supercoiled, whereas plasmids are small molecules, in large numbers of copies, and are coiled or supercoiled. During the extraction process, the aim is to denature the DNA and then renaturalize it. Because plasmids are small and supercoiled, they are renaturalized faster than the main DNA; the latter is trapped by protein complexes during renaturation, which makes it heavy and causes it to precipitate (Lodish, H. et al 2002; Sambrook J, and Russell DW, 2001).

Objectives

- Understand the plasmid DNA extraction protocol.

- Potentiate cognitive, social and verbal skills.

- Develop skills in the handling of laboratory instruments
  Materials

      Pseudomonas fluorescens strain
      Solution 1: 50mM glucose, lOmM EDTA, 25Mm and Tris HCl
      pH 8.0. Solution 2: 0.2N NaOH, 1%SDS (freshly prepared)
      Solution 3: To 60 mi of 3M sodium acetate, add 11.5 mi of glacial acetic acid
      and 28.5 mi of cold water.
-    Ethanol 95% and 70%
      Incubator

Microcentrifuge
Nutritional Broth with Micropipettes
Eppendorf tube rack.
Eppendorf tubes of 1.5 mL Paper towels
1000 pL Micropipettes Blue tips for the Micropipettes
- Ice

Methodology

For the extraction of plasmid DNA, the mini-prep technique of alkaline lysis described by Sambrook and Russell (2001) is used, the steps of which are the following with some modifications:

1.   Inc
ubate the bacteria in liquid medium Luria Brittany overnight at 37 ° C.

2. Take 3 mi of the previous culture and centrifuge for 5 minutes at 14000 r.p.m. and discard the supernatant.

3.   Resuspend the pellet in 100 pl of solution 1 (50mM glucose, 10mM EDTA, 25Mm and Tris HCI pH 8.0) (it must be cold) and incubate for 5 minutes at room temperature.

4. Add 200 pl of solution 2 (0.2N NaOH, 1%SDS) (cold and freshly prepared. Mix well by immersion and incubate for 5 minutes on ice.

5.   Add 150 pl of solution 3 (to 60 mi of 3M sodium acetate is added 11.5 mi

49

glacial acetic acid 28.5 mi of cold water) mix by immersion and incubate for 5 minutes on ice.

6. Centrifuge for 10 minutes, 4 or C, 14000rpm Carefully transfer the supernatant to another tube.

7.   Add 95% ethanol to the supernatant and incubate for 3 minutes at room temperature.

8. Centrifuge for 30 minutes and discard supernatant. Add 70% ethanol and centrifuge for 30 minutes, 4 °C, 14000 r.pm.
9.   Finally, suspend the pellet in 50 pl of TE (10 mM Tris HCl pH 8.0 and 1 mM EDTA pH 8.0) and store at -20°C until use.

Questionnaire

1. Explain the function of each of the components that are part of the solution?

2. What effect does solution 2 have on the DNA extraction procedure?

3. What is the purpose of solution 3 in the procedure?

4. Explain why ethanol is added at 100% and 70% cold?

Bibliography

- Curtís, H. and Barnes, N. Sue. 2000. Biology. Panamerican Medical Ed. 6th Edition. Madrid Spain.

- Lodish, H., Berk, EL, Zipurssky, S. L., Matsudaira, P., Baltimore, D. Damell, J. (2002). Cellular and Molecular Biology (Fourth Edition). Editorial Médica Panamericana. Madrid, Spain.

- Madigan, M., Martinko, J., Parker, Brock J. (2010) Brock, Biology of Microorganisms. 10th edition. Prentice Hall.

- Sambrook J, and Russell DW, (2001) Molecular Cloning: A laboratory manual, 3rd ed, Coid Spring Harbor Laboratory Press, New York.

Antagonistic potential of *Trichoderma harzianum*

By: Hugo Mauricio Jimenez M.

Introduction

Due to the high use of agrochemicals for pest and weed control, the environment is becoming quite polluted since these chemicals are xenobiotic compounds, i.e. synthesized by man, and their degradation in the environment can last up to 500 years.

For this reason, it is important to design other alternatives such as biological control of phytopathogens and pests, since due to its biological origin it does not alter the ecosystem, is easy to handle and low cost, which today has become an effective control strategy.

Biological control is a method of controlling pests, diseases and weeds using living organisms to control populations of plant pathogens.

The biological control when it works has many advantages among which we can highlight:

- Little or no harmful side effects towards other organisms including man.
- Pest resistance to biological control is very rare.
- Biological control is often long-term and permanent.
- The treatment with insecticides is significantly eliminated.
- The cost/benefit ratio is very favorable.
- Avoids secondary pests.
- There are no problems of intoxication.

From an economic point of view, an effective natural enemy (biological controller) is one that regulates the population density of a pest and keeps it at levels below the economic threshold established for a given crop.

Although a great diversity of natural enemy species have been used in a large number of biological control programs, species that have proven to be effective have certain characteristics in common that should be considered in planning and conducting new programs such as

- Adaptability to changes in the physical conditions of the environment
- High degree of specificity to a given target (pest to be controlled).
- High capacity of population growth with respect to its target (pest to be controlled).
- Synchronization with the phenology of the target (pest to be controlled) and ability to survive periods when the target (pest to be controlled) is absent
- Capable of modifying its action according to its own density and that of the target (pest to be controlled)

Before performing biological control field tests, it is a priority to perform *in vitro* studies of biological controllers with target organisms to analyze their antagonistic effect and predict how their action would be when used.

In vitro bioassays allow us to observe the beneficial or antagonistic effect of a chemical or biological factor on the growth of an organism. These bioassays are performed in the microbiology laboratory under controlled conditions and results are obtained in a short time.

Trichoderma harzianum has been characterized for being a good controller of phytopathogenic microfungi such as *Verticillium albo - atrum, Rhizoctonia solani, Sclerotium cepivorum* and *Fusarium oxysporum,* among others. It is widely used in agricultural biotechnology for its optimal growth in large scale production, and its easy application in field.

Trichoderma harzianum is a microfungus of the *Phylum Ascomycota,* which are a very diverse group, and are found in varied environments such as soil, salt water, fresh water and in all climates, present ecological applications as saprophytes and pathogens of plants and animals. These microfungi present a sexual phase known as telomorph by the formation of ascospores and an asexual phase known as anamorph by the formation of conidia.

These asexual phase microfungi are easily cultured *in vitro and* grow rapidly in liquid or solid culture media, which is why they are widely used in biotechnology for their optimal growth in bioprocesses.

Trichoderma harzianum is characterized by being a microfungus with an asexual phase due to the formation of a cruciform phialid and conidia formation of green grass color and its growth is abundant.

Objectives

- To perform antagonism tests against the phytopathogens *Verticillium albo - atrum* and *Sclerotium cepivorum* using the microfungus *Trichoderma harzianum.*

- Acquire skills in assembling in *vitro* antagonism tests.
- Develop skills in the handling of laboratory instruments

- Potentiate scientific skills.

Materials, reagents and equipment

- Microfungi *Trichoderma harzianum, Verticillium albo - atrum* and *Sclerotium cepivorum.*
- Petri dishes.
- Handle for microfungi.
- Erlenmeyer
- Potato Dextrose Agar Culture Media (PDA).
- Distilled water.

- Autoclave.
- Laminar Flow Chamber.
- Incubator.

Methodology

Preparation of Potato Dextrose Agar Culture Media (PDA)

The volume of PDA Culture Media for each petri dish is 25 mL, so it is necessary to make the calculations from the Oxoid Potato Dextrose Agar flask, weigh the amount, add it to an Erlenmeyer with the required volume of distilled water, cover with aluminum foil and reinforce with paper and masking tape, sterilize in the autoclave along with the petri dishes, once the sterilization is over serve the culture media in the petri dishes in the laminar flow chamber.

Antagonism Tests

1. In petri dishes with the potato dextrose agar (PDA) culture medium, sow with a mushroom handle a fragment of *Trichoderma harzianum and* confront it with the microfungus test to sow *Verticillium albo - atrum and Sclerotium cepivorum* respectively, as shown in Figure 3.

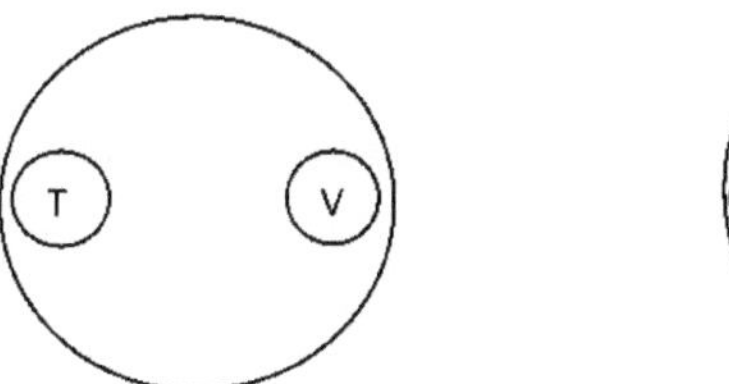

Figure 3. Sowing scheme, T: *Trichoderma harzianum,* V: *Verticillium albo - atrum,* **S:** *Sclerotium cepivorum.*

2. Leave to incubate at 25 °C for 9 days.

3. **Measurement of** mycelial **index** After the planting of the microfungi, measurements of mycelial growth corresponding to each antagonistic *in vitro* test are made every 3 days. To measure the mycelial index we used the equation proposed by Pérez et, taken from (Bonilla, 2005), **((MB-MA)/MB)*100%),** where MA is influenced mycelial growth and MB is free mycelial growth (See Figure 4).

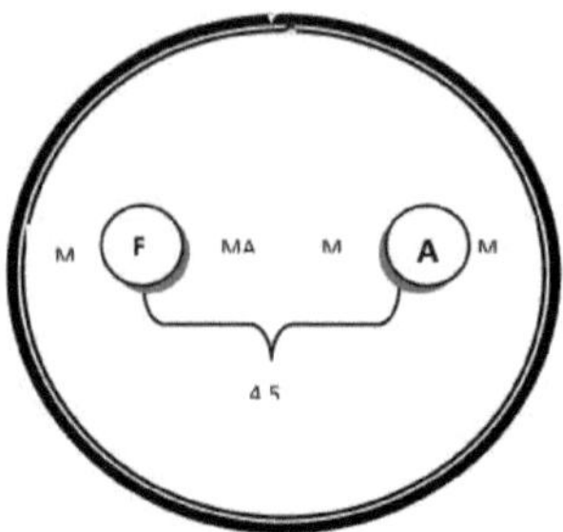

Figure 4. Scheme of the proficiency test, where FP is the phytopathogenic microfungus, A is *T. harzianum*. MA is influenced growth, MB is free growth. Taken from (Bonilla, 2005).

4. Observe the colonies, calculate the mycelial index of the colonies of the microfungi *Trichoderma harzianum, Verticillium albo - atrum* and *Sclerotium cepivorum.*

5. Also observe if there are structures between the colonies.
6. Conclude what was observed.

Questionnaire

1. See other *in vitro* antagonism tests between microfungi controllers and microfungi and/or phytopathogenic bacteria.

2. See *in vivo* tests for antagonism between control microfungi and microfungi and/or phytopathogenic bacteria.

3. See other species of microfungi used in biological control.

4. See species of bacteria used in biological control.

Bibliography

- Bonilla, A. 2005. Adaptive strategies of paramo plants and High Andean Forest in the Eastern Cordillera of Colombia. Bogotá. National University of Colombia, Science Faculty.

- Caicedo, V., 2014. Evaluation of the antagonistic effect of *Trichoderma harzianum* against microfungi and phytopathogenic bacteria of the Cepario of the Biotechnology Laboratory - UPN. Graduate work. Dept. Biology. National Pedagogical University. Colombia.

- Jimenez, H.M., 2007. Agromicrobiology Laboratory Guides. University of Pamplona. Faculty of Basic Sciences. Department of Microbiology. Colombia.

Cybergraphy

- Biological Control:
https://www.ecured.cu/Control_biologico Review
Date: May 14, 2019.

Biofertilizers for the improvement of the growth and development of the Soya.

By: Silvia Gómez Daza.

Introduction

Plants require nutrients from the air and soil for their growth and development; the richer the soil, the better they will grow and produce higher yields. However, if only one of the necessary nutrients is scarce, their growth and production decreases. Consequently, in agriculture, fertilizers are used to provide crops with nutrients to improve production.

A fertilizer is a chemical, organic or biological mixture used to enrich the soil with nutrients and to promote plant growth. For a product to be considered a fertilizer, it is indispensable that it be soluble and chemically available to the plant, since of the 18 nutritional elements considered essential for plants, 15 of them are taken in solution as ions. The chemical form in which the plant absorbs all the necessary nutrients for its correct development is the same regardless of the origin.

Chemical fertilizers are inorganic products obtained through chemical processes, elaborated in laboratories or factories, and are not very friendly to the environment; organic fertilizers are those produced from the decomposition of dead plant and animal remains, and biological or biofertilizers are products based on beneficial microorganisms in the soil, especially bacteria and/or fungi, which can live associated or in symbiosis with plants and naturally help their nutrition and growth, besides being soil improvers.

Within the biofertilizers there are different types: those that produce growth factors, phosphorus collectors and solubilizers, and nitrogen fixers. The microorganisms promoters of the vegetative development are those that during their metabolic activity produce and release regulatory substances of growth (auxins, cytokinins and ethylene) for the plant, examples *Trichoderma harzianum, Enterobacter aerogenes, Azotobacter sp and Bacillus mycoides* (González, H. and Fuentes, N. 2017).

Phosphorus collectors have the ability to increase the area of capture and absorption of nutrients mainly phosphorus through the roots of plants (mycorrhiza). Mycorrhizae are symbiotic or mutualistic associations of plant roots and fungi that allow the increase of the uptake rate of Phosphorus (P) and other nutrients such as Nitrogen (N), Iron (Fe) and Copper (Cu). There are two kinds of mycorrhizae: ectomycorrhizae and endomycorrhizae. In ectomycorrhizae the cells of the fungus form a large

56

size, with a slight penetration of the hyphae into the root tissue and endomycorrhizae are found mainly in forest-forming trees, especially conifers, beeches and oaks, and are more developed in boreal and temperate forests (Madigan, et al., 2010).

Organisms involved in phosphorus transformations (phosphorus solubilizers) in the soil include bacteria, fungi, chromista, protozoa and some nematodes. In general, soil microorganisms energize the P cycle through mineralization, immobilization and solubilization processes, which are related to their nutritional metabolism. The mechanisms used are: the production of organic acids, the production of protons (normally associated with the assimilation of NH4+ and/or respiratory processes), and the production of inorganic acids and C02- Within the genera of microorganisms used for solubilization are: *Pseudomonas putida, Micrococcus, Bacillus subtilis, Aspergillus niger,* among others (Patiño C. and Sanclemente O. 2014).

Nitrogen (N) fixing microorganisms have the ability to transform atmospheric nitrogen into ammonia and thus be able to supply it to crops. This process occurs through plant-bacteria symbiosis. One of the most interesting and important interactions is that between bacteria of the genus *Rhizobium and Bradyrhizobium* and legumes (soybeans, beans, clover, alfalfa peas etc.). The bacteria induce the formation of nodules in the roots within which the process of N fixation is carried out. Under normal conditions if the plant or the bacteria are alone the process does not occur, it is necessary to associate them. The plant provides the organic source of energy needed by the root nodule bacteria and the bacteria provides the fixed nitrogen for the development of the plant. So plants with nodules on their roots can grow in nitrogen-poor environments where others are unable to believe. In the field, *Rhizobium* is able to fix N only under controlled microaerophilic oxygen conditions, within the nodule the amounts of this compound is controlled by the leghemoglobin that serves as an "oxygen buffer" by binding to it; the formation of this protein is induced by the symbiotic interaction of these two organisms (Madigan, et al, 2010).

In symbiosis, the plant has the genetic information for symbiotic infection and nodulation; the role of the bacteria is to trigger the process. The stages of infection and nodule development include: a) chemotherapeutic attraction of the bacteria to the plant, b) bacterial fixation to root hairs, c) invasion of root hairs by the formation of a bacterial chain, d) bacterial development into root cells, e) formation of bacteroids within plant cells and development of the nitrogen-fixing state, and f) continuous cell division of the plant and bacteria as well as the formation of mature root (Madigan, et al., 2010).

In addition to the leguminous-rhizobic relationship, nitrogen-fixing symbiosis takes place between non-leguminous plants and other microorganisms. The water fern (*Azolla*) has the
characteristic of being associated with cyanobacteria in water bodies, especially Anabaena *{Anabaena azollae)* and thus fix atmospheric nitrogen. This cyanobacteria has the capacity to fix atmospheric nitrogen, reaching 1200 kg of fixed nitrogen per hectare per year under optimal conditions of temperature, soil and chemical composition of soil and water. For this reason, it is considered that *Azolla-Anabaena* can be a very important natural source of nitrogen within agriculture and has been used in rice and corn crops (Montaño M. 2005 and Aldás J., et al 2016).

Objectives

- Analyze the effects that fertilizers have on the development of soybeans.

- Potentiate cognitive and scientific skills.

- Understand the importance of biofertilizers

- Develop procedural skills.

Materials

- Soybeans (3-5 per treatment)
- Cotton
- Water
- Four containers with soil, one for each treatment.
    Treatment 1: control (water only)
    Treatment 2: chemical fertilizer
    Treatment 3: biological fertilizer : biofertilizer *(Rhizobiol* https://repository.agrosavia.co/handle/20.500.12324/20746)
    Treatment 4: organic fertilizer (mixture of eggshells, coffee beans and chicken manure)

Methodology

1. Take 3 to 5 seeds per treatment.

2. Depending on the treatment, perform the following procedure:

    2.1 For the first treatment only add water when you find the soil almost dry.

    2.2 For the second treatment, which is chemical fertilizer, add it when the plant have 4 true leaves and add water when you find the soil almost dry.

    2.3 For the third treatment which is with the biofertilizer (Rhizobiol) which is in liquid form based on symbiotic nitrogen-fixing bacteria specific to soybean cultivation do the following: place the contents of the biofertilizer in a clean container and add the seed, then mix until all the seed is well covered with the product. Then let it dry in the shade. Finally sow the seed immediately in the ground.

    2.4 For the fourth treatment, which is organic fertilizer, add it when the plant have 2 true leaves and add water when you find the soil almost dry.

NOTE: For treatments 1, 2 and 4, first put the seed to germinate in moistened cotton and then sow it in the ground.

3. Follow up each treatment every week for two and a half months; make the report where you include the following table for each treatment:

Treatment (tto)	Week			
Aspect to observe and describe	Number of sheets	Stem description	Color of the leaves	Stem size and thickness.
Seeds with water (tto 1)	59			
Seeds with chemical fertilizer (tto 2)				
Seeds with biofertilizer: *Rhizobiol* (tto -•3)				
Seeds with organic fertilizer (tto -4)				

Questionnaire

1. Write in two paragraphs what conclusions you can reach with the results obtained.

2. Compare your results with those of your colleagues and in two paragraphs comment on what conclusions you can reach and explain why.

3. Make a concept map with the topic of fertilizers.

4. Explain the importance of the use of biofertilizers.

Bibliography

- González, EL; Fuentes, N. (2017). Mechanism of action of five microorganisms plant growth promoters . Rev.Cieñe. Agr. 34(1): 17-31. doi: http://dx.d0i. org/10.22267/rcia. 173401.60.

- Madigan, M.,Martinko,J., Parker, BrockJ . (2010) Brock, Biology microorganisms. 10 edition. Prentice Hall.

- Montaño M. (2005) Study of the application of *Azolla Anabaena* as a biofertilizer in the cultivation of rice in the Ecuadorian Coast. Rev. Tecnol.18(1): 147-51

- Aldás J., Z. J., Cruz E., Villacís L, Pomboza P., León O. (2016) Biofertilizer effect of Azolla - Anabaena in maize (*Zea mays* L.) Fertilizer effect Azolla - Anabaena in maize (Zea mays L.) *JSelva Andina Biosph.* 4 (2): 109-115.

- Patiño, C., and Sanclemente, O. (2014). *Phosphorus-solubilizing microorganisms (MSF):*
a biotechnological alternative for sustainable agriculture. Magazine:
Entramado. Yol. N°2.UniversidadLibredeColombia   .   Recovered from   :
http://www.redalyc.org/pdf/2654/265433711018.pdf

- *Rhizobiol https://repository.agrosavia.co/handle/20.500.12324/20746*

Printed by Books on Demand GmbH, Norderstedt / Germany